No. 3. Serial. Price, 10 Cts.

THE

PULPIT AND ROSTRUM.

Sermons, Orations, Popular Lectures, &c.,

PHONOGRAPHICALLY REPORTED BY ANDREW J. GRAHAM, CHAS. B. COLLAR, AND FELIX G. FONTAINE.

LECTURE ON THE GREAT UNFINISHED PROBLEMS OF THE UNIVERSE.

BY

PROF. O. M. MITCHELL.

DELIVERED AT THE ACADEMY OF MUSIC, NEW YORK, JANUARY 29, 1859.

NEW YORK:
PUBLISHED BY E. D. BARKER,
No. 348 BROADWAY.

February 15th, 1859.

The Pulpit and Rostrum.

It is the object of this periodical to furnish, through the assistance of a corps of experienced phonographers, accurate reports, daguerreotypes as it were, of such sermons and lectures as are of the greatest public interest, and thus present to all, in a form suitable for preservation, specimens of the thought and style of our most celebrated public speakers. Thus will be secured many discourses of great interest and value, which otherwise would be lost to all but the immediate audience. We therefore offer the "Pulpit and Rostrum," in the hope that it will receive from the public a liberal support.

The Pulpit and Rostrum, No. 1, contains a *verbatim* report of the eloquent sermon of Rev. T. L. Cuyler, on Christian Recreation and Unchristian Amusement, delivered at Cooper Institute, Oct. 24, 1858. As this excellent discourse has called forth responses from several advocates of the drama, it possesses at this time special interest.

Pulpit and Rostrum, No. 2, contains full phonographic reports of *two* able and practical addresses—one by Rev. Henry Ward Beecher, and the other by James T. Brady, Esq.,—on Mental Culture for Women. These addresses make a powerful appeal in behalf of the working women of our country, especially those in large cities, and present facts deserving of universal attention.

Price per number, 10 cents ; twelve successive numbers, $1. For sale, and subscriptions received, by all booksellers.

E. D. BARKER, Publisher,

348 Broadway, New York.

THE GREAT

UNFINISHED PROBLEMS OF THE UNIVERSE,

WITH THE MOTIONS OF THE SUN AND THE PLANETS THROUGH SPACE, AND THE DETERMINATION OF THE CENTER OF THE STELLAR UNIVERSE.

A Lecture delivered Saturday Evening, January 29th, 1859, at the Academy of Music, New York, by Prof. O. M. Mitchell, of Cincinnati, for his own benefit, upon an invitation of the audience given on the occasion of the conclusion of his Course of Five Popular Lectures on Astronomy.

Previous to the Lecture, a series of Resolutions, with reference to the erection of an Astronomical Observatory in the Central Park in the City of New York, was offered by Prof. Loomis, and seconded in an eloquent speech by Prof. Davies, in the course of which a high tribute of praise was paid to Prof. MITCHELL. The Resolutions were unanimously adopted. Prof. MITCHELL was then introduced by the Hon. Luther Bradish, and spoke as follows:

I KNOW not how to answer in fitting terms the greeting of this night. The honorable and flattering allusion which has been made to me by my old preceptor and personal friend I can not respond to. His feelings of kindness and affection for an old pupil have carried him far beyond the just limits which should have restrained his remarks. I appreciate his motives, but I aspire not to the high eminence upon which, in the kindness of his heart, he has sought to place me.

I have been called hither by the invitation of a number of personal friends and strangers to speak in behalf of Science. I came humbly at the call, and was told that I was to address a multitude in this vast building; but had I known the responsibility which has been imposed upon me, I am confident I could not have mustered the courage necessary to have passed the thousand miles of interval which separates your city from my home. I am happy, however, that I am here. Notwithstanding stormy and tempestuous weather, I have been greeted night after night by your kind faces, until I have learned to feel that, in some sense at least, you are all my personal friends. I have further evidence of this friendship in the fact that you are here again to-night, at the termination of a long and perhaps tedious course of lectures, and I thank you one and all from the very depths of my heart for this manifestation of your good-will.

If I have succeeded in contributing in the slightest degree to the advancement of that great movement which has been auspiciously inaugurated to-night, I shall esteem it the proudest effort of my life; and if some biographical sketch shall ever mark to posterity the fact that I ever lived, upon the page that contains the record, I would point my children to that paragraph which says, "Your father was in the outset connected with this grand enterprise"—an enterprise which I trust is to eventuate in the erection of the noblest and proudest structure that has ever been reared upon the surface of the globe to the science of the stars. (Applause.)

Permit me, as a stranger, to say a word or two with reference to this movement.

It is utterly useless, Mr. President and Gentlemen of the Committee, for you or for the great city of New York to attempt anything less than the erection of the noblest Observatory in the world. I do not believe that the people of this city would be satisfied with anything less; and you have the power, you have the means, you have the ambition, you have the skill, to accomplish whatever you resolve to do.

When I stood, some fourteen years ago, in my own little city before a multitude like the one which I now have the honor of addressing, and there for the first time lifted my voice in behalf of the erection of a noble structure, whose chief ornament should be one of the grandest instruments that science and skill has ever produced, I ventured to make an appeal of this kind: The Old World looks with comparative contempt upon the profound ignorance and inertness of the New. They point to us and say, Yonder is activity, and strength, and power, and vigor, but it is all put forth to grasp the almighty dollar. And when I stood before that great assemblage and said, Let us rescue our country from the stain that is thus resting upon it—let us show to the crowned heads of Europe that free, independent, republican America can take the lead even in Science itself, the response to my appeal afforded the most gratifying evidence that in the end this grand object would be accomplished. What is the result? A short time after the commencement of the undertaking—and at that day there was scarcely an Observatory in our country—I visited Europe. I went to Munich, the great center for the construction of these mighty instruments, and there I stood in the presence of the successors of old Fraunhofer and Utzschneider. I said to them, "Your predecessors sold to the Emperor of Russia the great Equatorial Refractor." And why? Simply because they desired that their skill and handiwork, displayed in this masterpiece, should fall into the hands of some profound astronomer, and thus give them a world-wide reputation. "Sell to me," said I, "poor simple republican that I am—and yet one of the nobles of our land—this mighty refractor, equal to almost any other in the world, at cost, in like manner, and I will guarantee that in the next ten years you will get more orders from the United States than all the other countries of the world together." They would not make the sale on these terms, and yet during that time they have received more orders from this country than from all others, and we have built more Observatories and erected more magnificent instruments than all the world besides. Now, our scientific men stand on the same high platform with those of Europe. They hail us as brothers in this grand and noble crusade against the stars. We are moving on together—a solid phalanx; the watch-towers are rising all over the earth, and the grand cry is, Onward! It is echoed from Observatory to Observatory. The sentinel is everywhere posted, and do you not mean to post one on your rocky heights? (Applause.) I know you do.

I come now to the discussion of the subject appropriated to this closing lecture; and in doing so, I can not sever this one from the others, but must regard myself as addressing the same audience, and this lecture as a continuation simply of those which have already preceded it.

I am to speak of the Unfinished Problems of the Universe.

This would seem to imply that there are some which are finished; but I know of none such absolutely. I believe that we are now permitted to announce that the great law of universal gravitation reigns throughout our solar system with absolute command and power. I believe that we can, almost with certainty, announce that its dominion reaches to the fixed stars; and when this is uttered, I think that I have told you all the problems that are finished in the astronomical world.

When we come to the examination of our own system, when we come to inquire whether we have determined the actual and positive movements of the sun, whether we have reached to the precise and critical knowledge of the movements of any planet, whether we are able to predict with absolute precision the place of any one of these revolving worlds, I answer, it has not been done.

All we have accomplished is an approximation to perfection. We are moving on from year to year, and every year increases the perfection with which we are enabled to trace out the movements of these wandering worlds.

Let me exemplify this matter by reference to one single phenomenon. About two hundred years ago, one of those who devoted themselves to the examination of the revolving worlds—one of the followers of old Copernicus—thought he had sufficiently examined the movements of Mercury to predict the fact that it would cross the disc of the sun, and be seen upon the solar surface as a dark, round spot. His computations, however, were such that he felt he must give himself a limit of about five days. Think of it—a limit of five days!

Now, as the planet occupies but a few hours in crossing the disc of the sun, if within this time it should happen that the transit should occur in the night, the astronomer would of course lose the opportunity of verifying his prediction. He watched, therefore, during these days with an intensity which you can scarcely comprehend; and at last his eye was greeted, and his heart gladdened, by finding the planet, true to his prediction, upon the disc of the sun.

But another period of eighty or ninety years in the history of

Science rolls away. The astronomers of Paris are all deeply interested and excited with the approach of another of these transits of Mercury. Their computations were such that they believed they could rely upon them within—not five days—but five hours of the time. The sun was to rise with Mercury upon his disc. The morning arrived, and, armed with their telescopes, they were waiting to verify their computations; but the clouds intervened between them and the sun, and when he rose he was utterly and absolutely invisible. They waited and watched, hoping the clouds would break away and give them the long-coveted opportunity of verifying their computations; but the limit of time rolled away, and the clouds did not disappear. At length, one after another becoming weary with the watch, yielded up in despair, left his post, and entirely abandoned the observation. But one more doubtful of the computation than the others watched on. At last there came a little rift in the clouds. Through that chasm he hurriedly sent out his telescopic ray, and there, on the rim of the sun, clung the round, black disc of Mercury, telling him precisely within what limit of time their computations were in error.

Five hours was the limit required at that time. We come down to a later period. The Observatory of which I have had the honor of the direction, so far as the building, and the mounting of a single instrument were concerned, was completed in 1845.

In May of that year it was announced that Mercury would again cross the disc of the sun. It was the first observation I ever attempted to make. I had computed with all the delicacy in my power the exact moment when the dark planet would touch the brilliant rim of the sun. I had gone yet further, and computed the exact point on the tremendous circumference of the sun where the contact would take place—for remember, the power of this telescope is so great, that the sun swells out with such tremendous magnitude, as to literally and absolutely cover the whole heavens from horizon to horizon, could it all be taken into the field of vision at one view. The point of contact was brought within the field of vision of the telescope. The eventful day arrived, and the sun rose bright and glorious. Not a cloud stained the deep blue of the heavens. As the hours rolled by, and the time approached, there I was, with feelings such as you can not conceive, understand, or comprehend. My assistants were around me, ready with their chronomoters to mark the moment of contact. I hoped and believed that our tables and computations were so accurate that five minutes of time would be a sufficient limit, and five minutes before the appointed time I took my place at the great telescope. There I waited and waited, until it seemed as if an age had gone. I called out, "Surely the time is passed—what of the time?" "Only a single minute!" Second by second, only a minute had rolled away. It seemed as though hours had been sweeping slowly by. Again I took my watch and waited, until again it seemed as though an age had passed. "Surely," said I, "the time is gone." "No—another minute yet." At last I caught the black disc of the planet just impinging upon the bright rim of the sun—in the limits of a minute? No; but *within sixteen seconds of the computed time!* (Applause.)

You see, then, the possibility of advancement. This was not my work, but it was the work of another—Le Verrier. Le Verrier had taken up the movements of the planet Mercury, and with a power and precision of investigation never surpassed had corrected the previous tables, and reduced the theory within such limits that it had now become possible to make these delicate predictions. Do not imagine, however, that after your great Observatory shall have been erected, there is nothing to do. There is everything yet to do. Reduce these sixteen seconds down to the tenth part of a second of time. Cut it down, and when you have cut down all other errors in like manner and proportion, you will be able to fix the longitude and latitude of your ship at sea, bearing your merchandise and precious freight to all the markets of the habitable globe, and they will wing their way over the trackless deep in perfect and absolute safety.

I will now direct your attention to the subject specifically assigned for this evening—the great problem of the movement of the Stellar Universe. In the course of my preceding lectures I have already given you some idea of the train of investigation which has led astronomers to adopt the theory that our solar system is sweeping with tremendous velocity through space, and moving at such rate that it passes over one hundred and fifty-four millions of miles every year.

Up to the present time no one has ventured to say what the character of this motion is. We are moving toward a certain point, and that point is only approximately known. Are we moving in some mighty curve? Are we moving in some vast circle? Are we sweeping in some tremendous ellipse? or are we moving in a simple right line toward the point whither the sun is urging his flight? If the sun is indeed moving in any vast circumference, so soon as we can determine the fact, and get a portion of its mighty curve, so soon as we can get a portion sufficient to determine the plane in which it lies, then somewhere in that plane, in the depths of space, will be found the mighty center about which the sun and solar system are revolving.

Up to this time, however, we have no knowledge on the subject. All we can say is this, that if this (illustrating) be the direction in which the sun is moving, and perpendicular to this line we describe a plane entirely around the heavens, cutting from the solar sphere a circle, somewhere in that mighty circle will be found the center about which the sun is revolving.

Within a comparatively short time the attention of astronomers has been directed to an investigation with which this is specifically combined, and it is nothing more nor less than this grand question: Is there, in the whole starry heavens by which we are surrounded, any great central body, any mighty controlling orb, which holds a proportion to the bodies by which it is surrounded, such as our central sun holds to the planets which sweep around it? Looking at our own system, and supposing this was by possibility a sort of picture hung up in the heavens on a miniature scale, in order that there might be realized in the starry firmament with which we are allied another mightier system, of which all the stars should constitute the sweeping planets, and in the center of the whole some grand controlling orb, magnificent in its proportions, grand in the quantity of matter which it contains, vast in its outline and circumference, and sufficient to hold these mighty worlds and to produce harmonious and perfect movement throughout the Stellar Universe—*is* there such an orb existing in space?

I answer, there is not. Why? Because we are enabled by the telescope to penetrate space in every possible direction. Ah! —you may answer—but you can only bring into your telescope the light that comes from luminous bodies, and if this vast central orb is non-luminous, your telescope fails, and you can accomplish nothing; and when you state that such a body does not exist, you state what you do not know.

There is another method by which we may acquire a knowledge of the facts in this case. If it is true that this mighty orb exists in space somewhere, surrounded by all these glittering stars, even if it be opaque, and sends to us no light, if it has the attractive power which belongs to our own sun, and if it be energized by this mighty power of universal gravitation which holds these starry worlds in its grasp, then we are enabled by means of the telescope to detect that fact, because in the immediate vicinity of this central body the stars will sweep more rapidly under its gigantic power then those at a greater and still greater distance—just as the planets nearest to our sun revolve with greater velocity than those which are more remote.

Now we have examined the whole starry heavens, we have mapped out these heavens, and located these stars. We know where they were at the beginning of this century; we know where they are now. We know the amount of change which has taken place, and in case there was one region in which stars are more rapidly moving than in another, we have a sufficient knowledge of the heavens to detect this point in space. We are therefore enabled to pronounce that such a mighty central orb does not exist anywhere throughout the Universe of fixed stars with which we are allied.

There being no such body, you may of course conclude that there can be no revolution around a center. That does not follow. Let me tell you why.

A few years only have passed away since an astronomer commenced the examination of what is called "double stars." Sir William Herschel is again the pioneer in this field of investigation, and he tells us that when he began he gathered from all tho catalogues of which he had any knowledge a list of all the "double stars" then known. I think the list consisted of about five. Since that time it has been increasing with tremendous rapidity. He himself ran it up to hundreds; his son, who succeeded him, ran it up to thousands. After a while Struve, who had charge of the great refractor at Dorpat, gave his whole observing energy to this one department of the heavens, and the result has been that he has published a catalogue, in some sense, almost without number, of these double stars, which exist strewn richly throughout the regions of space.

Now, we find, after a rigorous examination of these double stars, that it is utterly impossible for us to suppose that they are optically united, that they are accidentally so located in space that they are so close together as to give the appearance of union; and when we come to apply what is called the calculus of probabilities, we find a limit within which this possible optical appearance may occur, and everything beyond or inside of this limit must be a physical union. The stars are not merely accidentally located in this way; they are combined, the one with the other, each energized by the power of gravitation, and the two revolving about their common center of gravity. Now, this announcement which I make, extraordinary as it may appear to those who have not hitherto investigated it, has been fully carried out and verified by observation. We trace these revolving suns in their orbits until, under the gaze of man, some of them have performed entire revolutions. Many others are far advanced. Astronomers have gone yet farther, and, applying the great law of gravitation and the laws of motion, have actually predicted their periods—have given us an ephemeris which should mark the place of these bodies in coming time, and these predictions have been verified, so that we have these revolving orbs scattered throughout the heavens; some rapidly sweeping through space in periods shorter than the periods or revolutions of our own planetary orbs; others rising in grandeur and magnificence until we find their periods reaching by possibility millions of years.

Let me call your attention to a single example. There is a quadruple star in the constellation Lyra—two double stars—the periods of which have been determined comparatively, and we find that one double set is revolving in this manner about the other—all of them sweeping through space and performing this mighty revolution in a period of not less than a million of our years. But you may ask me, how is it possible to decide such a question as this—how can it be done? First, we announce that these bodies are physically united, from the fact that they are all moving together in one common direction, with one equal velocity through space. I do not refer now to their movements or revolution about each other. I refer to a common proper motion, a sort of tie carrying these bodies off bodily together. It is utterly impossible that they should be carried off together unless they were physically united. They make up a mighty system, and when we come to measure the distance by which these bodies are severed, it is possible to determine roughly the period of revolution which must by necessity make up the vast time which is required for them to sweep entirely around. Thus we find that there is a diversity in the constitution of this universe, such as we find surrounding us everywhere upon the face of the planet that we inhabit. We may anticipate, therefore, schemes and systems rising one above another, each as diverse from the other as are the plants and animals that grace, dignify, and beautify the earth. So it is in the heavens. Here we have bodies of all possible kinds and characters.

If we take the telescope and look out upon the universes by which we are surrounded, we find them diversified in every possible way. Our own mighty Stellar System takes upon itself the form of a flat disc, which may be compared to a mighty ring breaking out into two branches, severed from each other, the interior with stars less densely populous than upon the exterior. But take the telescope and go beyond this; and here you find, coming out from the depths of space, universes of every possible shape and fashion; some of them assuming a globular form—and when we apply the highest possible penetrating power of the telescope—breaking into ten thousand brilliant stars, all crushed and condensed into one luminous, bright, and magnificent center. But look yet farther. Away yonder, in the distance, you behold a faint, hazy, nebulous ring of light, the interior almost entirely dark, but the exterior ring shaped and exhibiting to the eye, under the most powerful telescope, the fact that it may be resolved entirely into stars, producing a universe somewhat analogous to the one we inhabit. Go yet deeper into space, and there you will behold another universe—voluminous scrolls of light, glittering with beauty, flashing with splendor, and sweeping a curve of most extraordinary form and of most tremendous outlines. What is the meaning of all this? Nothing but the diversity with which the Almighty Architect has chosen to mark the superstructure by which we are surrounded. So that we may anticipate all the diversity that exists here on the earth and in the heavens beyond us in the system with which we are allied.

Is it then possible for us to find a center about which the whole Stellar Universe may be revolving? Admitting that there be no central orb there, admitting that there be no grand, central, controlling body, can we find a common center of gravity of the whole entire system? I doubt not that the time will come when this question will be answered in the affirmative. I trust the resolution of this great problem will be contributed to by your own Observatory with others that rear their summits toward the heavens, and from which the sentinel is looking out upon these deep, blue skies, marking the movements of these wondrous orbs. But a long time must roll away before we shall gather all the data necessary to give us the exact solution of this great problem. But a solution has been commenced; it has been attempted by one of the most distinguished astronomers of Europe, the successor of Struve, at the great Observatory at Dorpat—Maedler—who has distinguished himself in the astronomy of the double stars, and who, by his computations, examinations, and investigations, has placed himself on a level with the most distinguished men of the age.

I speak thus highly of Maedler because his theory is not now adopted by the best minds of the world. They are scarcely willing to accept it as yet. This was true with regard to the theory of Sir William Herschel, when he announced that there was a point toward which the solar system was sweeping in space, and he believed that he had found it. Years, scores of years rolled away before this was received, but it was a pioneer announcement—the announcement of a brave, bold, and daring mind, one that dared to speak, no matter whether the world listened or not; and now we see the result. Then this German astronomer, in the service of the Emperor of Russia, dared to put forth this grand conception of his; whether it be sustained or not, is a question which posterity has to resolve.

Let me give you some of the train of reasoning which he adopted in attempting to fasten the point about which the whole Stellar Universe is revolving, and our own sun among the number. First, then, the Herschels have revealed to us the figure of the great stellar stratum to which we belong. We know that the stars are condensed in a certain plane, which we call the Galactic Circle. They are more numerous there; they are nearer together there, and heavier, in some sense, when you come to take the mass within a given area, than you find in any other region. Now suppose this to be that plane. As we rise above it, toward the North, the stars grow fewer in number in a given space; they are more widely separated from each other, in some sense, and the stratum is comparatively shallow in that direction, as it is down below, toward the South. We are now enabled to determine the position of our own sun in this stratum, and we find it comparatively central in its location, and that we are nearer to the South than the North. Now, if this be true, we may anticipate that the center of gravity will lie toward the North. I have already announced that we have determined the direction of solar motion precisely. If we sweep a circle perpendicular to this plane round the whole heavens, somewhere in the region of this circle we may hope to find the center about which our own sun is revolving, and if we find this center, it is the common center of gravity of the entire scheme of stars.

Such was the nature of the research which first guided Maedler in his examinations. He began by looking at various large stars in the heavens. His approximate observations led him to the region of the constellation "Taurus." He first commenced by supposing that by possibility the brilliant star in the eye of the Bull —"Aldebaran"—might be the central sun, but a rigorous examination soon demonstrated that this could not be so. He then looked a little farther toward the South, and there he beheld that mighty and beautiful cluster of stars which we call the Pleiades.

Seven of them are visible to the naked eye; but when we turn the telescope upon this cluster, we find hundreds coming up to greet the vision of man, presenting one of the most beautiful and magnificent spectacles that is to be found in the whole heavens. Here is, then, a vast multitude of clustering worlds, and in the center of this cluster a bright and brilliant star, named by astronomers Alcyone. Maedler thought that by possibility this might be the center. How should he then undertake to verify the truth of this hypothesis? He began by a critical examination of what is called the *proper motion* of all the stars composing this cluster—all of them that had been mapped down; and it happened fortunately that this particular cluster had engaged the attention of Bessel, with his great heliometer. Many years before, he had fastened the places of some fifty or sixty with wonderful delicacy and precision, and by comparing Bessel's observations with those of other astronomers that preceded him, and of others that followed him, it became possible to determine the amount of proper motion belonging to each and every one of these stars. Now, when the proper motion is examined, it is found to be almost identical for every one of them. Here is a most remarkable fact. Suppose these stars not to be associated in any specific manner; suppose them to be grouped together by chance, if you please. Why should they, in consequence of the movement of our own sun through space, all of them appear to sweep away together? This is utterly and absolutely impossible in one sense, unless you suppose them all to be crowded and condensed together, so as to become, in some sense, a solitary body.

It is just as if you were sweeping along the line of a railway, and should see far off in the distance a little cluster of trees. By comparing their places with some more remote object, they might all appear to move together toward you. But suppose this cluster of trees should be expanded, separated, severed, and swept out to greater distances; then, you perceive the motions would be all different. Fixing your eye upon a distant object, one of these trees would move with a certain velocity, and another with a different velocity, and another with a still different velocity. And so in sweeping out the telescopic ray to this mighty cluster of stars in the Pleiades, they ought to appear to change; they ought to seem to sever, the one from the other, if it be only occasioned by the fact that they are located in a sort of line in this way, some near to us, some in the center, and some more remote. They do not thus exhibit themselves to the eye of man. Their proper motions are all the same.

Now, if Maedler adopted the idea or hypothesis that here was the center of gravity of the universe, he could then commence a train of reasoning to verify the hypothesis. If this be the center, then our own sun is sweeping around that center, and the stars on the hither side will appear to move in a certain direction; the stars beyond will appear to move in a certain other direction; the stars on the outside of the sun's mighty orbit in opposition to it will appear to move in a certain direction; and the stars that, so far as our own sun is concerned, happen to occupy that circle perpendicular to the line of the motion of the sun, will have a certain direction of motion.

Now, it had been shown in a paper of extraordinary interest and very profound investigation, that a large number of the conditions required to make this hypothesis the true one are verified by the examinations of the telescope. I do not pretend to indorse the theory of Maedler with reference to his central sun. If I did indorse it, it would amount simply to nothing at all, for he needs no indorsement of mine. But it is one of the great, unfinished problems of the universe which remains yet to be solved. Future generations are to take it up. Materials for its solution are to accumulate from generation to generation, and possibly from century to century. Nay, I know not but thousands of years will roll away before the slow movements of these far distant orbs shall so accumulate as to give us the data whereby the resolution may be absolutely accomplished. Bnt shall we fail to work because the end is far off? Had the old astronomer that once stood upon the watch-tower in Babylon, and there marked the coming of the dreaded eclipse, said: "I care not for this; this is the business of posterity; let posterity take care of itself; I will make no record;" and had, in succeeding ages, the sentinel in the watch-tower of the skies said, "I will retire from my post;

I have no concern with these matters, which can do me no good; it is nothing that I can do for the age in which I live"—where would we have been to-night? Shall we not do for those who are to follow us what has been done for us by our predecessors? Let us not shrink from the responsibility which comes down upon the age in which we live. The great and mighty problem of the universe has been given to the whole human family for its solution. Not by any clime, not by any age, not by any nation, not by any individual man or mind, however great or grand, has this wondrous solution been accomplished, but it is the problem of humanity; and it will last as long as humanity shall inhabit the globe on which we live and move. (Applause.)

I have been laboring myself to contribute my mite toward the resolution of this grand problem. Will you bear with me while I advert to some personalities in my own history, which are, in some sense, drawn from me by necessity, in connection with this discussion?

When the Observatory in Cincinnati was erected, when the debt which devolved upon me as an individual was finally paid off, the hope I had of sustaining the institution in connection with the college with which I was allied was, in a single hour, utterly destroyed forever. Fire seized the college-building, and in an hour all lay in ruins. The foundations of my hopes were swept away. What was then to be done? Give up the Observatory? That could not be done. I had no means; I had to depend upon my professorship. Everything I had in the world had been used up in accomplishing the building of this institution; and the only chance left me was the very one I am using to-night—to attempt to give popular lectures to sustain myself and the institution. And from that day to this I have sustained that Observatory by my lectures; and all the funds derived from the one I am giving to-night go at once into that treasury. (Applause.) I saw at a glance that I had no opportunity of attempting to carry forward a regular course of astronomical observations on a scale such as marks the movements of the great Observatories of the world. I could not come in competition with national institutions, endowed by the strong and bottomless purse of a government, imperial, kingly, or republican. I was but a solitary individual. I looked over the whole field of Science, and I finally resolved to enter the most forbidding field which presents itself to the mind of any one who devotes himself to the stars. It was nothing more nor less than to attempt to perfect the observations whereby we get the data for resolving all the problems by which we are surrounded, and to obtain greater rapidity and facility in making observations. I have been devoting myself to this one object for ten long years. What I have done I shall not speak of, except in its connection with the future movements of astronomical observations.

We have converted time into space; this was the first grand accomplishment. A second of time by the old method was marked out by the beats of a clock. When the observer desired to fasten the precise moment at which his star crossed the meridian wire of the telescope, fixing his eye upon the star, with his ear he took up the beat of the clock. This was the exact order of observation. He commences his count—"Five, six, seven, eight, nine;" and between "nine" and "ten" the star passes the meridian wire. He divides the space over which the star appears to pass in a second of time into ten equal parts, as nearly as he can, and enters in his note-book that the star passed the meridian wire at so many hours, so many minutes, so many seconds, and so many tenths. That was the old method. If the astronomer were called upon to mark the passage upon many wires, as is often done in a transit instrument, when he shall have obtained the passage of the first wire, he stops and enters it in the note-book. He must keep the count of the clock, and he must keep his eye upon the star; and his attention is divided between a variety of objects.

Now, by the new method, the clock records its own beats, takes care of itself; and the astronomer has nothing to do with it. An electro-magnet under the control of the pendulum of the clock (which by its motion, swinging backward and forward, moves a delicate wire upon an axis, so as to dip it at every swing into a cup of mercury, and close the circuit) brings a point down, and strikes a dot upon the disc revolving with uniform velocity to meet it; so that at the end of every second a dot is struck upon this disc; and thus, dot by dot, every second of time is formed into space.

Then, taking up the micrometer, we may cut the intervals between the dots into ten thousand parts; and thus we divide them down to any degree of exactitude we may demand.

When this great experiment was made in the outset, I attempted to unite this little piece of revolving wire, moving up and down, with the telescope, by some material sufficiently delicate and perfect to accomplish the result. I found it next to impossible to get any material which would answer the purpose. So delicate had the wire to be, that a single fiber or filament of silk, or a single human hair as fine as ever graced the head of beauteous maiden, was all too coarse for this purpose. It had not the requisite spring for such a delicate movement; and when this point dipped into the mercury it rebounded, and there were several touches instead of one. At length I went again to my old friend the spider, and asked him to aid me in this dilemma. I spun from him a web, which for three long years in every second of time was expanded and contracted, and performed the mighty service of uniting literally and absolutely the heavens with the earth. (Great applause.)

Now, what is the new method of observation? When the star enters the field of view, the observer, located at his transit instrument, has near him a magnetic key, such as belongs to all the telegraphic offices in your city. That key being struck, brings down a pen-point by the action of electro-magnetism. Here [illustrating] is the revolving disc; here the steel point; and when the key is touched, down comes the point, and striking upon the disc rebounds instantly, and the disc moves on uninterruptedly; and thus you have time, from second to second, converted into space upon the circumference of the disc. When one circumference of the disc is full, the disc moves itself on a little railway track just far enough to present a new circumference for another line of dots; and when the disc is filled, you have a perfect time-scale, absolute in its character, on which the clock, by automatic power, has recorded its own beats, and made a perfect record of itself. On that disc, by another magnet, another point is drawn down, and strikes, at the will of the observer, the precise moment at which he marks the transit of the star across his meridian wire; so that all he has to do is this: Take his place at the telescope, watch the coming of the star, pay no attention to the clock (for that takes care of itself), and at the exact instant at which his eye catches the bisection of the star by the wire, touch the key, and the record is made, and all is done. Thus you perceive that, by this new method, the astronomer is relieved from a large amount of intense responsibility resting upon him by the old method, wearing out his nerves, destroying his system, and rendering him, at the end of a certain time, incapable of continuing his observations. Another advantage gained, is that we may introduce as many wires as we please upon which to mark the transit of the stars, and thus reduce our observation to as great a degree of precision as we may desire. By the old method, remember, the observer is compelled to stop after the passage of one wire, and record the observation; and while he is doing that, the star is going on. By the new method, he has nothing to do but to touch a key, and the observation is recorded. Such is a rough outline of this new method of astronomical observation. So perfect is this method, that we read from the disc with the utmost possible facility; and we have conducted this examination in such a manner that now the thousandth part of a second is a quantity of time which we appreciate and employ every day.

If you will permit me, I will continue for a few moments longer to give you some of the details with regard to these new methods. Soon after the application of these new methods, it was manifest that we could, by the magnetic telegraph, determine the difference of longitude between two places with wonderful precision. Here is one of the greatest triumphs of modern science. When we reflect upon the results reached by the telegraphic communication between distant points in the determination of longitude, it seems positively as though modern science and skill paid no longer any attention to time; that it just crushed, crowded, and condensed a hundred years into a single hour! And it has been done. You go now and examine and determine the difference of longitude between the great Observatory of Paris and that of Greenwich, and you will find that by the telegraphic method, by these new means, we get better results in a single hour of one night than had been reached by all preceding time, although they had worked for two hundred years. So you perceive that in this particular department of Astronomy, in linking together the different observatories of the world, we have now an advantage that no old astronomer ever possessed. I myself have been somewhat engaged in this kind of work. The process is so exceedingly simple that I believe this entire audience will go with me and understand it without any difficulty.

Suppose the object were to determine the difference of longitude between this city and Philadelphia. The city which is farthest east will have a meridian such that the star will cross earlier than in the western one. Now, suppose the two observers are in telegraphic communication, and that there is a disc at each extremity receiving clock-beats. The observer in New York signalizes his friend in Philadelphia, and says, "The star is coming up to New York—look out!" and standing by the telegraph, the instant the star passes his meridian, he strikes the magnetic key, and the moment it is recorded on his disc it is recorded in Philadelphia. Then the Philadelphian waits until the star comes into his field of view, and he signalizes his friend in New York that the star is in the field of view; and the moment the transit occurs, he strikes the key, and the record is made. The interval of time between the two records is the difference of longitude. The process is perfectly simple; there is no sort of difficulty about it; you all comprehend it. (Applause.)

There are now two delicate questions yet to ask. The difference of longitude is actually obtained upon the supposition that all is perfect, and that this swift-winged messenger, the lightning, has flashed from one point to the other with infinite velocity. If it do not travel with infinite velocity, if it has lagged at all by the way, in communicating the messages, that amount of error will be entailed upon the result. Then it becomes necessary to investigate the great problem—With what velocity does the electric current flash along the wire?

I have had the opportunity of investigating this problem myself. I worked at it with an intensity of interest which you can scarcely comprehend, alone as I was—buried, in some sense, in a wilderness—with no one to sympathize with me in these strange investigations. I succeeded, however, in surrounding myself with a number of assistants who took a deep interest in this matter. I secured a telegraphic communication of wire entirely around from Cincinnati to Pittsburg and back again, in order to determine whether electro-magnetism accomplished the circuit of 607 miles of wire instantaneously, or whether it took some time. My disc was prepared, and the clock-beats were being received upon it. I arranged in such a manner that two pens should record upon a metallic disc by steel points, by the most delicate dots imaginable, the time for the passage of the electric current. Then I prepared inside the Observatory a short circuit of six or eight feet of wire, and to that battery I gave the identical intensity which belonged to the battery for the long circuit. I then arranged in such a manner that I could interchange these two points with each other, making one move with the long circuit and the other with the short, at pleasure. Having arranged the whole apparatus, I watched with the deepest interest to see whether the clock-beats, as recorded by the two pens upon the disc, would fall at the same moment of time, or whether an interval would exist which the eye or the ear could detect. But when the pens fell, it wanted a keener and sharper ear than mine to detect any difference. I then looked to see whether the dots struck were in a straight line radiating from the center of the disc; but with the most rigorous examination I could make, I could discover no difference. I was compelled to restrain my curiosity until the night should pass and the daylight come again. Then, with an instrument constructed for the purpose of measuring the thousandth part of a second, I measured the interval between the two dots—those struck by the *long*, and the others by the *short* circuit. I found invariably the same result in more than a thousand observations.

How much time do you suppose it took the electric current to flash around the 607 miles of wire? I give you my own result. I divided a second into one thousand equal parts; and that journey was performed in twenty-one of those parts—in twenty-one thousandths of one second of time. (Applause.)

But there is another difficulty in this process, growing out of the fact that the observers who make the record, who send the signals, who make the observations, may be different men. They must be different men. (Laughter.) But they may differ literally and absolutely in their physiological organization; and we must take this into account, and investigate it and determine it, before we can reach absolute correctness. Can this be done? Let me explain this matter; it is perfectly simple.

Suppose this entire audience were here now in utter and absolute darkness, and each held in his hand a magnetic key ready to touch it at the moment that I flash upon you a flash of lightning, and on one of these revolving discs each and every one of you could record the precise moment at which you perceive the coming of this flash. Strange as it may appear, it would be found that each one makes a different record; and if you try it a thousand times, each of you will find out the peculiar organization inside—a sort of curious spiritual daguerreotype of what you are on the inside—how you are made up—and you will all be different; but the most remarkable fact is, that you will differ from what you were an hour ago! (Applause.) That is not an uncommon thing in these days, to differ in a single hour. (Laughter.)

This fact of constitution is what we call personality. When the difference between two individuals has been determined, we call it a Personal Equation. (Laughter.) What is the meaning of this? Suppose the Philadelphia observer is so constituted that he sees the star cross the wire a tenth of a second earlier than his correspondent in New York; then we must allow for that difference of a tenth of a second. You may say, "What is the use of talking about the tenth of a second?—that is nothing." It is nearly an age in astronomy! One tenth of a second of time! Why, converted into space, it is one and a half seconds of an arc! "What is that?" I tell you it is five-fold greater than the parallax of 61 Cygni, which we have been compelled to determine in order to ascertain the distance of those revolving suns. Think you that we are to neglect the tenth part of a second? I tell you if we can not measure below the tenth part of a second, and drive our errors out of the tenths into the hundredths, and out of the hundredths possibly into the thousandths, we may as well stop observing, for we have already rough data enough! (Applause.)

Now, is it possible to determine the "personality" of each individual? I have finally succeeded in making a piece of machinery of delicate character which enables me to measure the "person ality" of any individual, in a single minute of time, by his making certain observations. Permit me to explain this. Suppose this disc was revolving with uniform velocity. Now, on the edge of the revolving disc I place say ten vertical wires about an inch in length, so that they may be readily distinguished. That disc revolving, immediately behind any local, fixed position there will be a dark line just as broad as the diameter of the wire; and you yonder, looking at the moving body, with a key in your hand make the circuit of a battery such that the electro-magnet shall record the moment at which you see the passage. If that same wire is made, by dipping into a cup of mercury, to make a connection with the battery and record its own passage, the difference between its record and your own will be your peculiarity; and your personality thus comes out beyond any question. (Laughter.) In other language, if I can make the star which crosses the wire in the field of the telescope record the moment of its passage, and you do the same, the difference between the two records will be your personal equation.

This has been accomplished; but you can form no conception of the amount of time which has been devoted to it. I have studied upon this alone for three long years. This and other experiments which I have made in this connection have amounted to more than one hundred thousand. I know that they are nearly all to be thrown away. They are but the center of the mighty arch and dome which we have been attempting to rear; and when the superstructure is up and the keystone placed, we shall knock away the center, and there the work shall stand as long as time shall last. (Applause.)

I can not detain you to speak of the other departments of astronomical observation, to which I have given much attention. Suffice it to say, we are now recording the places of the stars in our Observatory with a rapidity and accuracy I think hitherto unheard of. The observer takes his place at the telescope. An assistant is located in such a manner as to read the difference of north polar distance between any assumed standard star and the stars whose places are required, and just as fast as the stars can come into the field of view we find it possible to mark their places, and fix their position, and catalogue their magnitudes and peculiarities. Thus we are sweeping a zone of five degrees in width with an accuracy and precision equal to that of micrometric work. How many stars, think you, we are thus enabled to mark down in a single minute of time? I have taken that group of the Pleiades, and in five minutes I have fastened the places of from thirty to forty stars. In a single hour, in the richer portions of the Milky Way, in a zone of a single degree in width, I have recorded the places of more than one hundred stars. I hope, therefore, that the time is coming when the stars can not take refuge in their numbers and distance, and defy the power of man to dislodge them from the high concave in which they are entrenched. We shall grapple with them there; we shall hunt them down; we shall record their places; we shall number them as they come out from the depths of heaven under the penetrating gaze of the great telescopic eye which man has turned toward the stellar sphere! Will you do your part in this grand work? Are you ready to begin? Are you prepared to give a helping hand to the sentinel who gives his time, his talent, and all that he has on earth, to this grand and magnificent investigation?

Now, my friends, I must close this long course of lectures. We have passed from planet to planet, from sun to sun, from system to system. We have reached beyond the limits of this mighty stellar cluster with which we are allied. We have found other island universes sweeping through space. The great unfinished problem still remains—Whence came this universe? Have all these stars which glitter in the heavens been shining from all eternity? Has our globe been rolling around the sun for ceaseless ages? Whence, whence this magnificent Architecture, whose architraves rise in splendor before us in every direction? Is it all the work of chance! I answer, No. It is not the work of chance. Who shall reveal to us the true cosmogony of the universe by which we are surrounded! Is it the work of an Omnipotent Architect? If so, who is this August Being? Go with me to-night, in imagination, and stand with old Paul, the great

Apostle, upon Mars' Hill, and there look around you as he did. Here rises that magnificent building, the Parthenon, sacred to Minerva, the Goddess of Wisdom. There towers her colossal statue, rising in its majesty above the city of which she was the guardian—the first object to catch the rays of the *rising*, and the last to be kissed by the rays of the *setting*, sun. There are the temples of all the gods; and there are the shrines of every divinity. And yet I tell you these gods and these divinities, though created under the inspiring fire of poetic fancy and Greek imagination, never reared this stupendous Structure by which we are surrounded. The Olympic Jove never built these heavens. The wisdom of Minerva never organized these magnificent systems. I say with St. Paul: "Oh, Athenians, in all things I find you too superstitious; for, in passing along your streets, I find an altar inscribed, To the Unknown God—Him whom ye ignorantly worship; and this is the God I declare unto you—the God that made heaven and earth, who dwells not in temples made with hands."

No, here is the temple of our Divinity. Around us and above us rise Sun and System, Cluster and Universe. And I doubt not that in every region of this vast Empire of God, hymns of praise and anthems of glory are rising and reverberating from Sun to Sun and from System to System—heard by Omnipotence alone across immensity and through eternity! (Great applause.)

THE PLANETARY AND STELLAR WORLDS;

A POPULAR EXPOSITION OF THE GREAT DISCOVERIES AND THEORIES OF MODERN ASTRONOMY. IN A SERIES OF TEN LECTURES.

BY PROFESSOR O. M. MITCHEL.

1 vol. 12mo., cloth. Price, $1 25.

"For a practical, comprehensive exposition of the principles of astronomy, as they are now understood, no better work can be found. Written in a glowing style, the great principles and facts of the science are stated in that popular language which every reader can understand, and which presents the author's thoughts in the clearest manner. For the use of schools, and for private reading, we think it will win its way at once to a universal popularity. It is illustrated by a series of admirable engravings, which add, of course, incomparably to the excellence and utility of the work."—*New York Evangelist.*

"In itself considered, this is one of the most interesting, entertaining, instructive and valuable works that we have perused in many a day. We have read it with feelings glowing more and more with the finishing of every page; and have longed to put it into the hand of every inhabitant of earth whose soul leaps after the systems of worlds and suns which circle above him, and onward to immortality and heaven."—*Spectator.*

"The work gives a most admirable popular exposition of the great discoveries and theories of modern astronomy, and cannot fail to be universally read with the greatest profit and delight. We commend it to attention and favor."—*Courier and Enquirer.*

POEMS AND PROSE WRITINGS.

BY RICHARD HENRY DANA.

2 vols., 12mo., cloth. Price $2 50.

Mr. Dana has written far too little to do justice to his fine powers; but what little he has written is of the noblest and purest order. In all the highest qualities of the poet he is not exceeded by any writer in our country; and if lasting fame is to rest upon the merit, rather than upon the amount of what an author produces, he will maintain his place among the foremost of our poets. The prose sketches which form a part of these volumes are among the most exquisite productions of their class within the whole compass of our literature.

"Mr. Dana's writings are addressed to readers of thought, sensibility, and experience. By tenderness, by force, in purity, the poet paints the world, treading in safety the dizziest verge of passion, through all things, honorable to all men; the just style resolving all perplexities, a rich instruction and solace in these volumes to the young and old who are to come hereafter."—*Literary World.*

"Mr. Dana is evidently a close observer of nature, and therefore his thoughts are original and fresh."—*True Democrat.*

"In addition to the Poems and Prose Writings included in the former edition of his works, they contain some short, practical pieces, and a number of reviews and essays contributed to different periodicals, some of them as much as thirty years since, and now republished for the first time. As the expression of the inmost soul, these writings bear a strong stamp of originality."—*N. Y. Tribune.*

Copies of these Books sent by Mail, postpaid, for price remitted to

CHAS. SCRIBNER,

124 Grand St., New York.

"Cooper emphatically belongs to the nation."—WASHINGTON IRVING.

"While the love of country continues, his memory will exist in the hearts of the people."—DANIEL WEBSTER.

"The creations of his genius shall survive through centuries to come."—W. C. BRYANT.

The Works of the Greatest American Novelist

ILLUSTRATED BY THE

GREATEST AMERICAN ARTIST!

ILLUSTRATED EDITION

OF

COOPER'S NOVELS,

WITH DRAWINGS ON STEEL AND WOOD,

BY F. O. C. DARLEY.

The Most Magnificent Enterprise of the Day.

The Publishers have the pleasure of announcing the publication of an entirely new edition of COOPER'S NOVELS, containing the latest improvements and revisions, printed on superfine cream tinted and calendered paper, from the most perfectly formed type, in a large Crown Octavo page, elegantly bound in embossed Cloth, with bevelled edges, and Illustrated with designs on wood, and vignettes on steel, executed with BANK NOTE finish, in Line and Etching, from drawings by DARLEY, that great master of design. In elegance, beauty and artistic finish this series will be the most perfect publication ever issued in this country.

To Americans the name of Cooper is endeared as being the creator of a distinctly original class of Fiction, so strictly national in its character, and so finished in the vividness of its execution as to have marked its author as the proud compeer of the highest names in the list of European Novelists; and in view of the wide popularity of these novels, the publishers have prepared this splendid NATIONAL EDITION, worthy of the author and his fame, and illustrated by an artist of congenial genius, who for years has made the pages of Cooper a study, and whose drawings have been conceived in a spirit and with a breadth and power worthy of his author. As a National Literary Enterprise combining the productions of a great national author with those of a great national artist, and issued with commensurate elegance, this edition of Cooper, the publishers believe, will receive the support and approval of the American public.

The Publication to commence February, 1859, a volume issued monthly; each volume will contain a Novel complete, crown octavo, averaging 500 pages each; Published by subscription, at $1 50 per vol.

W. A. TOWNSEND & Co., Publishers.

377 BROADWAY, N. Y.

[TURN OVER.]

"His writings are instinct with the spirit of nationality."—W. H. PRESCOTT.

ASTRONOMICAL SERIES.

Mattison's Primary Astronomy.

For Schools and Families; adapted to the capacity of Youth, and il trated by nearly Two Hundred Engravings. By HIRAM MATTISON, A.M., Professor of tural Philosophy and Astronomy in the Falley Seminary; Author of "High School tronomy," etc.

168 Pages Large 18mo. Price 40 Cents.

This is an admirable first book in Astronomy. It is arranged in the form of questions answers, with numerous explanatory notes. By the aid of its very numerous and finely cuted illustrations, every thing is made plain, and most interesting, even to younger pupil

Mattison's High School Astronomy;

In which Descriptive, Physical and Practical are combined, with espec reference to the wants of Academies and Seminaries of Learning. By HIRAM MATTIS A.M., etc., etc.

240 Pages 12mo. Price 75 Cents.

This work is divided into three parts. After an introduction, which consists of prelimin Observations and Definitions, and occupies twenty pages, Part First is devoted to the S System—the Sun, Planets, Comets, Eclipses, Tides, etc.; Part Second relates to the Side Heavens—the Fixed Stars, Constellations, Clusters, and Nebulæ; and Part Third to Pract Astronomy—the Structure and Use of Instruments, Refraction, Parallax, etc. This dep ment, so seldom introduced into text-books for schools, will be found especially interest and valuable.

Besides embracing all the late discoveries in astronomy, under a strictly philosoph classification, the work is thoroughly illustrated by the introduction of diagrams into pages, in connection with the text; and the adaptation throughout to the use of the bla board, during recitation, cannot fail to be appreciated by every practical teacher.

Burritt's Geography of the Heavens,

And Class-Book of Astronomy. Accompanied by a Celestial Atlas. By E JAH H. BURRITT, A.M. Greatly Enlarged, Revised, and Illustrated, by H. MATTISON, A.

Geography 338 Pages 12mo: Atlas, Imperial 4to, with 9 Colored Maps. Price $1 50.

This has long been the standard text-book in Astronomy in American Colleges and Hi Schools. Its popularity is evinced in a sale of 250,000 copies.

As its title imports, it is designed to be to the starry heavens what geography is to the ear Both teachers and pupils have found that Class Books and Maps are as indispensable to t department of knowledge, as in that of Geography; and that an artificial globe is just as po a substitute in one case as in the other. Instead of the globe, and a few balls strung up wires, Mr. Burritt's book and Atlas introduces the pupil at once to the grand Orrery of t Heavens, and makes him acquainted with the names and positions of the bodies which co pose it. He learns to *locate* and to *classify* his astronomical knowledge as he does his g graphical; and experience has proved that a child of ten years will trace out all the Constel tions that are visible in the heavens, and name the principal stars in each, as readily as he w learn the boundaries of the States from a map, and name the cities they contain.

It has just been entirely re-stereotyped, new maps having been added to the chart, a many improvements having been made, which render the work, as it stands, very compl in all its details. The matter has been thoroughly assorted, new topical and suggestive qu tions having been prepared throughout, and numbered to correspond with the paragraphs which they relate. A complete list of Telescopic Objects, in each constellation, has been serted, which will greatly enhance the value of the work to all classes of students. Two e tirely New Maps have been introduced into the Atlas, containing views of eighty differe objects.

Published by MASON BROTHERS,

NEW YORK.

www.ingramcontent.com/pod-product-compliance
Lightning Source LLC
LaVergne TN
LVHW011132110826
845150LV00008B/2303

* 9 7 8 1 4 1 8 1 9 4 3 1 4 *